WHO PUT PEOPLE ON EARTH?

The True Origin of Humanity

Timothy Aldred

Also by Timothy Aldred

6 Steps to Religious Freedom:
Learn How to Reclaim Your Mind From Catholic Church
History,
Think For Yourself and Enjoy Your True Religious
Freedom

Bamboozled! Besieged By Lies, Man Never a Sinner:
How World Leaders Use Religion to Control the Populace

The Dark Side of Religion Revealed in Ten Pages:
A Companion to Bamboozled! Besieged By Lies, Man
Never a Sinner

The Dark Side of the Publishing Industry:
What Big Publishers Don't Want You to Know

Sexuality: Creation of a Blemished Society

TABLE OF CONTENTS

Preface..1

Chapter 1: WHO PUT PEOPLE ON EARTH?........7

Chapter 2: HOMO SAPIENS SAPIENS: A SLAVE
MADE TO ORDER..11

Chapter 3: THE ONLY COHERENT AND
INTELLIGENT ACCOUNT.......................................18

Epilogue..25

Credits and Citations..28

About the Author..29

Sneak Peek: 6 Steps to Religious Freedom............30

PREFACE

I was born in nineteen forty-six. And why do I have to say this?

Factually speaking, the Romans of Italy rose to power and changed the geographical calendar, where I was born. This calendar was once named the Julian calendar and it was later renamed the Gregorian calendar. It was used to facilitate the criminal and fraudulent innovation of Jesus Christ, and put into law. It introduced in Latin the term A. D. Anno Domini, and 'In the year of our Lord' in the English language.

In brief, I will explain the basic reason people are confused in their view of the world around them. They have grown up in a mutated and uncivilized culture. They will never understand unless the information that has been kept obscure from them is revealed. The citations will show anyone with an average reading ability that anyone can unravel the biggest religious scam ever, which has turned most people born in the Jesus Christ era into unnecessary wonderers. This is my effort to explain the incredibly evil act that occurred on earth (which destroyed the original human perception of self), and to explain how the forces of nature work.

Rome illegally invaded Israel in 63 B.C.E., oppressed and killed many of its citizens, seized their wealth, and destroyed Judaism, their religion. The behavior of the Roman regime was a clear indication that it wanted a religion of its own, to become earth's sovereign ruler, and in essence to become what is known today as God Almighty itself. They spent about one hundred years in Israel, leaving only after they accomplished the establishment of Christianity.

[See: Whose Land? A History of the Peoples of Palestine by James Parkes. The following is a public instrument, used to corrupt the people of the world.]

The Gregorian calendar now in use was introduced in 1582 by Pope Gregory XIII (1502-85), as the corrected form of the earlier Julian calendar. Pope Gregory did much to increase papal authority and to establish the temporal independence of the papacy. He sent St. Augustine on a missionary expedition to Britain in 596 A.D.

For people who care about knowing the different world situations that affect our life, we were not present in the past, so beginning in the kindergarten classroom, it's a simple matter of depending on educational sources to inform us. It is funny that after some people learn a language, and the rules of etymology, they turn into erroneous dictators. They begin to make up their own rules about what it is to form an intelligent procedural jurisdiction—which is unfortunate for those who choose to take that route of chaos.

I insist on following coherent linguistic principles. This is the luminous cognitive energy that allows for the equitable mitigation of all issues regarding the records of ancient times, now referred to as the Paleolithic Age vs. A.D. Anno Domini.

It is *crucial* that all people understand the deep paralyzing effect of the A.D. belief syndrome. This is because millions have been born and have died, such as my own parents, believing in the allegorical and fictional world, taken as a whole life of reality—when it is just a camouflage.

The corruption of our *original* human religious civility caused an insane viral effect, as people's natural sense of reasoning went out of control, from an over two-millennium reign of the Roman Gregorian Anno Domini calendar. Some may ask how the highjacking affected us. This is a plausible question, of course! The purpose of this effort is to explain how the oldest civilized ancestral records reveal to us how they were created. I will show a clear contrast of how the Roman calendar vehicle of Anno Domini adversely affected the human race in perpetuity, and devastated commonsensical reasoning.

Here is a caveat for you. Bear in mind the reality of the name Jesus Christ—it's fledgling and offered in the New Testament as a spurious revelation. Based on that fact, the word Jesus is always the allegory of the man. Pilate ordered him to be put to death by crucifixion. The historic facts show that Jesus did not exist in any way, shape, or form, but he is a fictitious being in the fictional cast of an allegorical world created by Rome, the Italian mafia. Always remember to ask: what was Rome doing in Israel to begin with; having dominion over a commonwealth of people?

Evidence for Jesus: The Anno Domini Dating System

Posted on January 1, 2009 by Kazz

It is true that today much of the world uses the Gregorian calendar with the Anno Domini dating system, and this fact is often brought up by Christians as evidence for the existence of Jesus, but this "evidence" would hardly even be worth refuting if it weren't so commonly used.

Most people don't know much about the history of our calendar, which is not surprising since it is not often important in our lives. However, when it is brought up in this way it becomes necessary to explore it, and it's actually even more interesting than might be expected.

This is the basic premise:

We use a dating system based on the birth of Jesus.

Therefore, Jesus existed.

To examine the truth-value of this claim, we must break it down further:

AD, which stands for "Anno Domini" means "in the year of our Lord". It did not spring into existence in 1 AD, or even 50 or 100 AD. The Anno Domini dating system was created in what we now call 525 AD by a monk named Dionysius Exiguus. This was more than

500 years after the monk believed Jesus to have been born.

To make matters worse, we know that Dionysius made mistakes in his dating system. This is a well established fact, and combined with the irreconcilable conflicts in Biblical accounts that attempt to date the birth of Jesus, it leaves us with little doubt that the Anno Domini dating system is not in fact accurately based on the year that Jesus was born.

About 200 years later, another monk and historian, who is now called Saint Bede, came up with "ante uero incarnationis dominicae tempus" ("the time before the Lord's true incarnation"). This is what we refer to as "Before Christ" or BC.

It should be clear why some monks would wish to create a dating system based on the year they believed their messiah to have been born, and it is not evidence for a historical Jesus, only for monks who believed in one.

Even in Bede's day, over 700 years after the time Dionysius set for the birth of Jesus, this dating system was not popular. It wasn't until later in the 8th century that it began to spread slowly across Europe after its endorsement by Emperor Charlemagne. Despite all of this, it took nearly 700 more years to cover even Western Europe, finally reaching Portugal in 1422. Even today, it is not used universally.

If the whole world, or even just people in the vicinity of Judea, had immediately started to date things based on the same BC/AD system we have today, it would be a strong piece of circumstantial evidence for the existence of Jesus. Even if they had started

*using the dating system within 50 years, it would have
been impressive. "*

In the world today, many errant views about
human origin have spread. This is because of the
problem discussed above and this is a confusion that
has prevented millions of people from knowing
mankind's beginnings on earth.

Categorically, going by the existence of early
civilization, such discussion does not permit the
existence of two accounts of human origins. With the
occurrence of the first civilized ancestry, all
subsequent aberrational accounts of mankind, in
every religious book, disappeared when we know
200,000 years of African historic residency, and of
the 100,000 years when people migrated to continents
and smaller land space. Rome had only forced people
to accept Jesus Christ by its enforcement of the
Crusader and the Inquisition. But they have not
managed to write anything other than the allegory
confabulated; not one word that resembles a plausible
attempt to report the truth.

CHAPTER 1

WHO PUT PEOPLE ON THE EARTH?

To prevent my thoughts from running wildly without purpose, I have decided to apply a strict boundary around what I say. Neither you nor I will be able to breach, the principles of the process that leads to coherent results about who put us on earth, and about what commonsense records tell us to accept as truth.

In deciding on genuine issues, we were born and became conscious of what we are encouraged to adopt as a credible standard of proof that serves as the highest confidence for having belief and faith in things. Historians say we are to look for five elements in geography that establishes the viability of a relevant whole idea: 1) location, 2) place, 3) human and environmental interaction, 4) movement, and 5) regions.

And in general history, like in the case I am presenting here, of human cosmogony and our origination in Africa—intelligence calls for six elements to satisfy the trust and reliance that wise people look for in assessing a piece of history. They are: 1) The world in spatial terms, like the name of streets, 2) place and region, with a special meaning, 3) physical system, noted for certain cattle or prone to a natural type of disaster, 4) human system; how boundary lines are determined and why people settle in certain places as opposed to others, 5) environment and society, the people and their natural surroundings if it influences the way people live, 6) the use of the geography, how war in one country affects a neighboring one.

Okay, that was to assist folks who may suffer from the fear of reading history with confidence.

What do you know about human cosmogony, the recorded thoughts, speech, and behavior of prehistoric times? Well globally, the consensus among people is that mankind originated in Africa. That agreement brings together those who believe in the biblical revelation of the Garden of Eden and the creation of man, and those who believe in the Paleolithic Age. For example, what happens, when Allah and God Almighty Holy Ghost brethren, gather in one place for a symposium on the subject of man's initial first day on earth. I see people in our true history making use of the African real estate, in all of the elements I pointed out above. Those things offer grounds for a plausible argument, saying people lived in Africa for a much longer time and generated parochial and anthropic human accounts further into the past than the people born yesterday, as it were. The ancestors were on the ground hundreds of thousands of years prior to their time. It is impossible for the bible to generate energy, forming truth from a sheer allegorical character called God and a six-day creation of heaven and earth. What all the A.D. religions have to rely when vying for recognition is far less than wet paper, to bear up human activities that occurred within a geographical area worth believing in. When one understands the word God in biblical terms, they no longer treat the bible as a divine book. Once etymology reveals the correct use of belief and faith, God and the six-day creation disappears. You see it is invalid and illogical.

I have no premise on which to present a competitive challenge to the evangelical biblical God of the allegorical six-day creation. So let me move on to presenting the only authentic knowledge of human origins. Let it be clear that humans are directly descended from actual, intelligent, bodily beings, not by an unknown spirit, as in the bible. Good history can only be identified by natural, wise, and prudent people.

So I ask for your attention as we pay respect to the truth of people on earth, and upon the strongest of the civilizations in history. There has not been the presence of any ghosts or God making anything. Since the advent of human consciousness, there were always the great Goddesses and Gods. It is on this historic rock of gods that we have knowledge of our true existence in the Abzu of Africa. From the earliest, to the last intelligent civilizations such as Sargon and Hammurabi, all had an awareness of living beings known as gods.

With the foundation in place, it's now time to fully address how we people got to be on dry land on earth. I heretofore share with you what the predecessor told us of how the embryonic phase came about in Africa. The account is coming to you based on Dr. Samuel Noah Kramer's extensive work in Mesopotamia, modern-day Iraq.

Thereupon I pass along to you in narrative form, Zecharia Sitchin's collaboration with Dr. Kramer's Assyriology and Sumerian cuneiform work.

CHAPTER 2

HOMO SAPIENS SAPIENS: A SLAVE MADE TO ORDER

The brainwashing of the A.D. bible created a split from the supreme knowledge of highly intelligent Homo sapiens sapiens, to a very limited state of mind. It produced the kind of thinking that accepts ideas like fallen angels marrying human women; God as an inscrutable spirit that made people who turned bad, etc.

After 2014 years, it is time that we woke up. The bible refuses to tell people they are hybrids of intelligent Homo sapiens—why? I am giving you the reason herein. However, for an individual to understand actual historic reality he must reject the allegory of scripture. Let us tighten up our consciousness of human intelligence.

Adam and Eve of the bible is a garbled allegory that has been extracted from our natural human history. Folks just need to know that the bible has been used as a camouflage to keep people from knowing literal history. It is proven that once you have accepted the stupidity of God's six-day creation, your mind may never return to its natural state. Today the world is partly full of people having their minds spoiled by anti-historic information. It is very frustrating when you have to defend real history against ignorant, unprecedented spiritual ideas that consist of a contrived, imaginary world.

Speaking from historic records, Sitchin wrote, *"The 'days' of the Book of Genesis are literally twenty-four periods and not eras or phases. The sequence in the bible is, as previous chapters should have made clear, a description of evolution that is in accord with modern science. The innumerable problems arise when the Creationists insist that we, Mankind, Homo sapiens sapiens, were created instantaneously and without evolutionary predecessors by 'God.' They simply used the allegorical statement in the bible that read…"And the Lord God formed Man of the dust of the ground, and breathed into his nostrils the breath of life, and Man became a living soul."*

This is obviously a Creationist error; taught in childhood in Sunday school, and found in chapter 2, verse 7 of the Book of Genesis. According to the King James English version, that errant thinking is what the Creationist zealots strongly hold as God's truth. They have not yet understood that the authors of the bible were simply borrowing from older historic accounts. This argument is according to historic information.

Were you to learn the Hebrew text, which is, after all the original historic record, you would discover that, first of all, the creative act is attributed to the Elohim–a plural term that at least should be translated as 'Gods' not 'God.' Secondly, you would become aware that the quoted verse also explains why 'Adam' was created. "For there was no Adam to till the land." These are two important—and unsettling—hints to who had created Man and why.

Then, of course, there's another problem, that of another (and prior) version of the creation of Man, in Genesis 1: 26—27. First, according to the King James Version, "God said Let us make man in our image, after our likeness;" then the suggestion was carried out: "And God created man in his own image, in the image of God created He him; male and female created He them." Take a further look and you will see that the biblical account confused the historic basis from which they drew their knowing in chapter 2. According to this, "Adam" was alone until God provided him with a female counterpart, created from Adam's rib. You see, the suggestion for Man's creation comes from a plural entity addressing a plural audience saying, "Let us make an Adam in our image and after our likeness."

What those who believe in the bible must ask is, what is *really* going on here? As both Orientalists and Bible scholars now know, what went on was the compilers of the Book of Genesis compiled, edited and summarized much earlier and more considerably detailed texts that were first written in Sumer. Those texts, reviewed and extensively quoted in Sitchin's book, *The 12th Planet*, with all sources listed, relegate the creation of Man to the Anunnaki. We learn from such long texts as Atra Hasis, when the rank-and-file astronauts came to earth for gold and ended up embroiled a mutiny. The backbreaking work in the gold mines, in southeast Africa had become unbearable. Enlil, their commander in-chief, summoned the ruler of Nibiru, his father Anu, to an Assembly of the Great Anunnaki and demanded harsh punishment for his rebellious crew. But Anu was more understanding.

"What are we accusing them of?" he asked as he heard the complaints of the mutineers.

"Their work was heavy, their distress was much!"

Was there no other way to obtain the gold? He wondered out loud.

Yes, said his other son Enki (Enlil's half-brother and rival), the brilliant chief of the Anunnaki. It is possible to relieve the Anunnaki of the unbearable toil by having someone else take over the difficult work. Let a Primitive Worker be created! The idea appealed to the assembled Anunnaki. The more they discussed it, the clearer their clamor grew for such a Primitive Worker, an *Adamu* to take over the workload. But, they wondered, how could you create a being intelligent enough to use tools and to follow orders?

How was the creation or "bringing forth," of the Primitive to be achieved? Was it, indeed, a feasible undertaking? A Sumerian text has immortalized the answer given by Enki to the incredulously assembled Anunnaki, who saw in the creation of an *Adamu* the solution to their unbearable toil:

The creature whose name you uttered....

IT EXISTS!

All you have to do, he added, is to

Bind upon it the image of the gods.

The irresistible problem that we face here in getting people to comprehend what is being said in the foregoing is the failure of *cognition*. Strangely, many don't even know they have any. Before I take this further, have a nice cup or your favorite drink: coffee, tea, or hot chocolate, while I set the tone for greater understanding:

COGNITION: *The mental process or faculty by which knowledge is acquired; that which comes to be known, as through perception, reasoning, or intuition and knowledge.*

The puzzle of Man's creation is now explained, and undoubtedly, it removes the conflict between Evolution and Creationism.

The Anunnaki Elohim did not create Man from nothing.

The being was already there on earth, the product of evolution. All that was needed to upgrade it to the required level of ability and intelligence was to "bind upon it the image of the gods," the image of the Elohim themselves. In simple terms, let us call the "creature" that existed then as Apeman/Apewomen. According to the ancient account, the chief scientist, Enki, knew how to "bind" upon the existing creatures, the "image" — the inner, genetic makeup — of the Anunnaki. In others words, he knew how to upgrade the Apewoma/Apeman through genetic manipulation, by just jumping the proverbial gun on evolution, and this was the process that brought "Man"—Homo sapiens—into existence.

SCIENTIFIC AMERICAN SAYS: COURTESY OF BERNHARD HAUBOLD

Humans, chimpanzees, gorillas, orangutans and their extinct ancestors form a family of organisms known as the Hominidae. Researchers generally agree that among the living animals in this group, humans are most closely related to chimpanzees, judging from comparisons of anatomy and genetics.

If life is the result of "descent with modification," as Charles Darwin put it, we can try to represent its history as a kind of family tree derived from these morphological and genetic characteristics. The apex of such a tree shows organisms that are alive today. The nodes of the tree denote the common ancestors of all the apices connected to that node. Biologists refer to such nodes as the last common ancestor of a group of organisms, and all apices that connect to a particular node form a clade.

There are two major classes of evidence that allow us to estimate how old a particular clade is: fossil data and comparative data from living organisms. Fossils are conceptually easy to interpret. Once the age of the fossil is determined (using radiocarbon or thermoluminescence dating techniques, for example), we then know that an ancestor of the organism in question existed at least that long ago. There are, however, few good fossils available compared with the vast biodiversity around us. Thus, researchers also consider comparative data. We all know that siblings are more similar to each other than are cousins, which reflects the fact that siblings have a more recent common ancestor (parents) than do cousins (grandparents). Analogously, the greater similarity between humans and chimps than between humans and plants is taken as evidence that the last common ancestor of humans and chimps is far more recent than the last common ancestor of humans and plants. Similarity, in this context, it refers to morphological features such as eyes and skeletal structure.

A common question among modern religious people is: *Who created heaven and earth? Who wakes me up in the morning?* If you grew up reading *actual* world history, as I have been discussing, the knowledge of human life would be as universally known as the very well marketed bible. Now it is obvious that the bible extracted all of its allegorical ideas of human creation from the official preexisting historic information.

CHAPTER 3

THE ONLY COHERENT AND
INTELLIGENT ACCOUNT

The well-explained birth of wise Homo Sapiens Sapiens is the archived information told by renowned civilizations—Akkadian, Sumerian and Babylonian just to name a few of them. Their account are the same as many organisms from one organ. In this case, the example of one organ means goddesses and gods; the only source that is able to produce the level of people's present intelligence. No credible record shows that any non-flesh entity indeed created people. It bars the baseless, inauthentic bible from challenging or competing with Paleolithic history.

From the University of Pennsylvania Dr. Samuel Noah Kramer tells of the trip he took to Mesopotamia, today's Iraq. It was to undertake Near Eastern studies deciphering cuneiform tablets of the late Bronze Age; dating back to circa 1300 B.C. Kremer devoted his life to the deciphering of cuneiform writing. He speaks of the contribution to the restoration and deciphering of the tablets bearing written cuneiform language.

Readers of my book *Bamboozled! Besieged by Lies, Man Never a Sinner* get further details therein. I would be remiss not to mention that Kramer left the institute of doctoral study in Near Eastern historic studies. You may also Google online to know more about Dr. Kramer, and the library also carries his books. I told you about Kramer because he is the leading authority on this profound area of history. Now another man name Zecharia Sitchin, championed the past civilizations cosmogony as well, and I have some of his book in my possession. I cite from his work pertinent issues to adequately satisfy readers' intellectual quest for this level of knowledge.

From Sitchin's book entitled *Genesis Revisited*, on the front cover is a question on which I would love to share my observation with you before going further:

Staggering new evidence of an astonishing science that flourished in mankind's distant past! Is modern science catching up with ancient knowledge?

In chapter nine of this book, the subject is "The Mother Called Eve." The context is one that displaced the allegorical presentation, persistent and commonly used by Christendom to convey a wrong sense of the world's known civil human history. It explains that Genesis authors deliberately used the process of transliteration to confabulate. This was meant to simply replace historic facts with pure fantasy. In that sense, we are today reading the Genesis creation story and are digesting in essence gross falsehood instead of good history—daytime became nighttime, what was the truth became a lie, and so on.

Therefore, only the energy of positive information can remove the deep-seated belief in the Genesis message. Those who believe cannot see the lies, because belief has no energy to bring realization and the means to have knowledge.

Sitchin: *Reading the biblical tale of Eve. It was the great Sumerologist Samuel Noah Kramer, who first pointed out that her name, which meant in Hebrew "She who has life," and the tale of her origin from Adam's rib in all probability stemmed from Sumerian play on the word TI, which meant both "life" and "rib."*

Some other original or double meanings in the creation tales, and more, can be gleaned about "Eve" and her origins from comparisons to the bible tales with the Sumerian texts and an analysis of Sumerian terminology.

Further, Sitchin writes: *The genetic manipulations we have seen were conducted by Enki and Ninti in a special facility called, in the Akkadian versions, Bit Shimti—"House where the wind of life is breathed in." This meaning conveys a pretty accurate idea of what the purpose of the specialized structure, a laboratory, was. But here we have to invite into the discussion the Sumerian penchant for word play, thereby throwing fresh light on the source of the tale of Adam's rib, the use of clay, and the "breath of life."*

You can read the book to get more of the details. It explains the complexity of the life-giving process in this segment, with Sitchin's interpretations.

In the absence of the original Sumerian version from which the compilers of Genesis might have obtained their data, one cannot be sure whether they had chosen the "rib" interpretation because it was conveyed by both IM and TI, or because it gave them an opening for making a social statement in the ensuing verse:

And Yahweh Elohim caused a deep sleep

Upon the Adam, and he slept.

And He took one of his ribs

And close up the flesh in its place.

And Yahweh Elohim constructed of the rib

Which He had taken from the Adam a woman,

And He brought her to the Adam.

And the Adam said,

"This is now bone of my bones,

Flesh of my flesh."

Therefore is the being called Ish-sha ["Woman"]

Because out of Ish ["Man"] was this one taken.

Therefore doth a man leave his father and mother

And shall cleave unto his wife

To become as one flesh.

Sitchin explicitly points out, in detail, the conflict between the original extant civilized account, and the subsequent A.D. biblical allegorical writing. He said that the confusion must be resolved, or no one will know the original situation, which showed that woman and man were created simultaneously, not as the bible wrongly copied. The following is the illustration that explains the original act:

In order to resolve this seeming confusion, the sequence of creating the Earthlings must be borne in mind. First, the male lulu, "mixed one" was perfected. Then the fertilized eggs of Apewoman, were bathed and mixed with the blood serum and sperm of a young Anunnaki, divided into batches and placed in a "mold," where they acquired either male or female characteristics. Re-implanted in the wombs of Birth Goddesses, the embryos produced seven males and seven females each time. But these "mixed ones" were hybrids, which could not procreate (as mules cannot). To get more of them, the process had to be repeated over and over again.

My intent is to present the readers, with the most informative historic information people have frequently asked of me. Of course, it would be wonderful, to end the religious confusion among people—I am just trying to help.

The account of Sumer by Sitchin continues:

At some point, it became apparent that this way of obtaining serfs was not good enough. A way had to be found to get more of these humans without imposing the pregnancies and deliveries on female Anunnaki. That way was a second genetic manipulation by Enki and Ninti, giving Adam the ability to procreate on his own. To be able to have offspring, Adam had to mate with a fully compatible female. How and why she was brought into being is the story of the Rib and of the Garden of Eden.

But we do know nowadays, thanks to modern science, that sexuality and the ability to procreate lie in human chromosomes. Each person's cell contains twenty-three pairs—in the case of women a pair of X chromosomes, and in the case of the men one X and one Y chromosome. The key to reproduction thus lies in the fusion of the two single sets of chromosomes. If their number and genetic code differ, they will not combine and the beings will not procreate. Since both female and male Primitive Workers already existed, their sterility was not due to the lack of X or Y-chromosomes.

The need for a bone—the bible stresses that Eve was 'bone of the bones' of Adam—suggests that there was a need to overcome some immunological rejection by the female Primitive Workers of the males' sperms. The operation carried out by the Elohim overcame this problem. Adam and Eve discovered their sexuality, having acquired "knowing"—a biblical term that connoted sex for the purpose of procreation ("And Adam knew Eve his wife and she conceived and gave birth to Cain). Eve, as the tale of the two of them in the Garden of Eden relates, was thenceforth able to become pregnant by Adam, receiving from the deity a blessing combined with a curse:

"In suffering shalt thou bear children."

With that, 'The Adam' Elohim has become as one of us. He was granted "knowing." Homo sapiens were able to procreate and multiply on their own! But though he was given a good measure of the genetic makeup of the Anunnaki, who made Man in their image and after their likeness, even in this respect of procreation, one genetic trait was not transmitted— the longevity of the Anunnaki. Of the fruit of the 'Tree of Life,' partaking of which would have made Man live as long as the Anunnaki, he was not even to taste. This point is clearly spelled out in the Sumerian tale of Adapa, the Perfect Man created by Enki:

Wide understanding he perfected for him....

Wisdom he had given him....

To him he had given Knowing

Eternal life he had not given him

EPILOGUE

It is fruitless to think that a hibernating spirit the bible called God, created mankind to have them live and behave wickedly throughout eternity. Why should one think that an omniscient God never wanted a world of peace, but one of swords and distress? Why would a perfect God make humans with glands to secrete energy that control thought and behavior, to want sex more than anything else on earth, or an insatiable desire to kill others?

To think along those lines is to accept that a fictional God who kills people himself, and one who ordained evil before the foundation of his world is holy and blameless, and that the Serpent who gave Eve and Adam knowledge, kills no one except those who God tells it to bite. In the face of this, there are still many people who fail to accept actual history, a flawless account of human existence, while believing in the baseless non-existent biblical message.

This work is designed to answer those questions that people are still asking about who created man. Hopefully, everyone will see that I have brought the ultimate historic information. If one thinks that they are not bound by it, then they are free to demonstrate what they know to be the legitimate source. For me, the following contains the essence of naturalness, cognition, and logic:

> *The Anunnaki Elohim was not a freak of nature but the result of deliberate experiments by Enki and Ninti. This is obvious from the Sumerian text. The text describes how the two came up with a being that had neither male nor female organs, a man who could not hold back his urine, a women incapable of bearing children, and creatures with numerous other defects. Finally, with a touch of mischief in her challenging announcement, Ninti is recorded to have said:*

How good or bad is man's body?

As my heart prompts me,

I can make its fate good or bad.

Having reached the stage where genetic manipulation was perfected to enable the resulting body's good or bad aspects, the two felt that they could master the final challenge—to mix the genes of hominids, Apemen; not with those of other Earthly creatures, but with the genes of the Anunnaki themselves. Using all the knowledge that they had amassed, the two Elohim set to manipulate and speed up the process of Evolution...Called upon to perform the task of "fashioning servants for the gods" – to bring to pass a great work of wisdom," in the words of the ancient texts—Enki gave Ninti the following instruction:

Mix to a core the clay
From the Basement of the Earth,
Just above the Abzu,
And shape it into the form of a core.
I shall provide, knowing young Anunnaki
Who will bring the clay to the right condition.

I hope this booklet has sufficiently wet your beak for more information. If so, read on!

CREDITS AND CITATIONS

Aldred, Timothy. *Bamboozled! Besieged By Lies, Man Never a Sinner: How World Leaders Use Religion to Control the Populace.* North Charleston: CreateSpace, 2011. Print.

Haubold, Bernhard. "How closely related are humans to apes and other animals? How do scientists measure that? Are humans related to plants at all?" Scientific American. October 23, 2000. Accessed March 3, 2014. http://www.scientificamerican.com/article/how-closely-related-are-h/.

Kazz. "Evidence For Jesus: The Anno Domini Dating System." Better Than Faith. January 1, 2009. Accessed March 3, 2014. http://www.betterthanfaith.com/articles/evidence-for-jesus-the-anno-domini-dating-system.

Kramer, Dr. Samuel Noah. *History Begins at Sumer: Thirty-Nine "Firsts" in Man's Recorded History.* Philadelphia: University of Pennsylvania Press, 1981. Print.

Parks, James. *Whose Land? A History of the Peoples of Palestine.* Middlesex: Penguin Books Ltd, 1949. Print.

Sitchin, Zecharia. *Genesis Revisited.* New York: Avon Books, An Imprint of Harper Collins *Publishers*, 1990. Print.

About the Author

At the tender age of seven, Timothy Aldred accepted Jesus Christ as his Lord and Savior. After more than fifty years in the dark, he discovered the light of mankind's true origins. Today his life is dedicated to sharing the truth as passionately as he once shared the lie.

Timothy can be contacted via his Website:
http://www.timothyaldred.com

SNEAK PEEK:

6 Steps to Religious Freedom:

Learn How to Reclaim Your Mind From Catholic Church History, Think For Yourself and Enjoy Your True Religious Freedom

By Timothy Aldred

PREFACE

If you're someone who trusts in the definition of words more than you do in your own reasoning, you could be wiser than you think.

I just learned about that a few years ago. You see, I was born and raised by Christian parents. In that church situation, I was conditioned to perceive life and history in a certain way, without even being aware that it was happening.

It all started with the things I learned in the church services: the words God, Devil and Jesus Christ. And so for the next fifty years, I saw things colored with God and the characters of the bible.

I even used to wonder why so many different churches were around, but now I see it as the way the idea of God works. Satan was also real to me, and whenever I felt negatively, it was all his fault.

Honestly, when I look back on those years, I see how unfruitful my life had been. I see how I clamored for fulfillment, but was never able to realize anything of substance.

So I designed this little six-step booklet to assist anyone who may be walking in my chapel shoes. For better digestion of the content, I suggest the use of a journal. There are some questions at the end of each step, please answer them honestly in your journal. It is important that you take the time to quietly reflect on your feelings, write down your present beliefs, and then compare them with the information I shared with you in that step.

The goal here with these six steps is to ignite your imagination, to illuminate the meaning of the words, so that your current mindset is clear to you and held up for scrutiny and observation. The most important objective is to notice the contrast between what I say and your present understanding of today's preaching.

You will find herein that there is zero tolerance for the false use of the words 'belief' and 'faith,' or for establishing things that are based on speculation and assumption. The reason for rejecting that kind of usage is because the definitions of belief and faith offer no basis for one to lay an authentic foundation regarding any issue. At this time, you are to take note of your responsibility as a reader, to observe *etymology* for word definitions.

I have based this guide on official world history, and the information it gives about our civilization's first religion. The exercise questions have been fashioned to weigh the factors between the information that pre-existed with our ancestors, in contrast of modern-day religious practices.

STEP # 1

Know and Understand Your True Self

It may seem a little odd for me to say that I got to know my real self when I was about fifty years old, but the reason I did not know myself sooner, was simply because, as I mentioned, I grew up with Christian parents who did not know *themselves*.

I think it is factual to say that no one can truly know his- or herself, unless they get to know the ancient historic truth concerning human life. And equally no one can know the reality of the world, unless they learn to read and are able to identify things for exactly what they are.

The religion of our modern time operates on the inappropriate use of belief and faith. That makes people psychologically blind, and keeps the things they need to know hidden away, thereby creating false or hereditary assumptions, and many a dilemma in life. The way to break free of the false teachings we learn in Sunday schools is by learning to follow the *meaning of words*.

It is important to remember that that's what keeps us connected to the order of relevance, logic, and reality. No one should just believe in what people say, unless it is everyday triviality, so that if what he or she say fails, you can easily recover from any loss.

As a rule of procedure, we must all put knowledge first in everything we do, unlike what the bible teaches, which is to put trust in God, even when no one has ever identified the object the bible calls God. So we are much better off putting our confidence in what words tell us.

Words are our best friends.

There are many things one should know to keep oneself from becoming foolish; some of them will be covered here. I am sure from here on out you will move ahead with your life, discovering the others on your own, because your "eye" will now be open.

There are certain things we must move out of our way, because they often sabotage progress. They ferment confusion and disallow the flow of challenge—and I am willing to tell you about these things.

It is very important that you see yourself as a product that belongs to Earth. And that recorded history that *pre-dates* all modern-day religious texts, shows that you are an upgraded species, called Homo sapiens sapiens. That means you are a hybrid, taxonomically genetic, manipulated being, that was created to breathe, sleep, eat, and walk. Understand that you were born with a computerized programming in your brain—to allow your body to evolve.

The process of common evolution happens each day, because of the earth's rotation and revolution.

Those motions directly affect our aging, and since that is the case, we know that the earth is the boss of our coming into the world.

Whatever happens to us after we are born is made possible by Earth's energy; the system which controls how long we live and when we will eventually die. It is very important to have a basic understanding of how life exists, because it acts as a shield and protects you from the BC/AD timeline, a religious delusion.

Everything I talk about in this guide is well thought out, and is designed to be followed for logical benefit. When these directions are followed, we develop the power to secrete a hormonal psychic power; this in turn increases our cognitive command. The knowledge we need is one that keeps us depending on *logic*, which only steers us towards the natural life inside of our bodies.

We get the natural truth about humanity from B.C.E. antiquity, the knowledge given by goddesses and gods. It is all about two elements of basic education, which orients and gives intellectual balance to you. There is no unaccountable, spurious God in ancient history, no so-called revelation for a place in heaven, or a burning hell.

End of Preview

Purchase a copy of 6 Steps to Religious Freedom in print or e-book format anywhere books are sold!

www.ingramcontent.com/pod-product-compliance
Lightning Source LLC
Chambersburg PA
CBHW071548170526
45166CB00004B/1593